Understanding the Elements of the Periodic Table™

THE NITROGEN ELEMENTS

Nitrogen, Phosphorus, Arsenic, Antimony, Bismuth

Greg Roza

New York

Published in 2010 by The Rosen Publishing Group, Inc.
29 East 21st Street, New York, NY 10010

First Edition

Library of Congress Cataloging-in-Publication Data

Roza, Greg.
The nitrogen elements: nitrogen, phosphorus, arsenic, antimony, bismuth / Greg Roza.
p. cm.—(Understanding the elements of the periodic table)
Includes bibliographical references and index.
ISBN-13: 978-1-4358-5335-5 (library binding)
1. Nitrogen—Popular works. 2. Nitrogen compounds—Popular works.
3. Periodic law—Popular works. I. Title.
QD181.N1R69 2009
546'.71—dc22
2008054258

Manufactured in the United States of America

On the cover: The atomic symbols and structures of nitrogen, phosphorus, arsenic, antimony, and bismuth.

Contents

Introduction

The chemical elements are the forms of matter that make up everything in the universe. They include materials like oxygen (O), carbon (C), iron (Fe), and gold (Au). Sometimes, elements are found in pure form, but often they are found combined in substances called compounds.

The elements are organized and displayed on a chart called the periodic table. This table makes it easier for scientists and students to find information about the elements. It is broken down into columns called groups and rows called periods. Each group contains elements that share similar traits.

The fifteenth column of the periodic table, known as group 15, contains five elements: nitrogen (N), phosphorus (P), arsenic (As), antimony (Sb), and bismuth (Bi). This group is often called the nitrogen group.

The nitrogen group was once called the pnictogen group. This comes from a Greek word that means "to choke." The origin of this term comes from the fact that pure nitrogen can cause asphyxiation (the medical term for suffocation) when breathed in because it does not contain the oxygen necessary to support life.

Did you know that nitrogen compounds can explode? Did you know that we need nitrogen compounds to live? How can one element be so

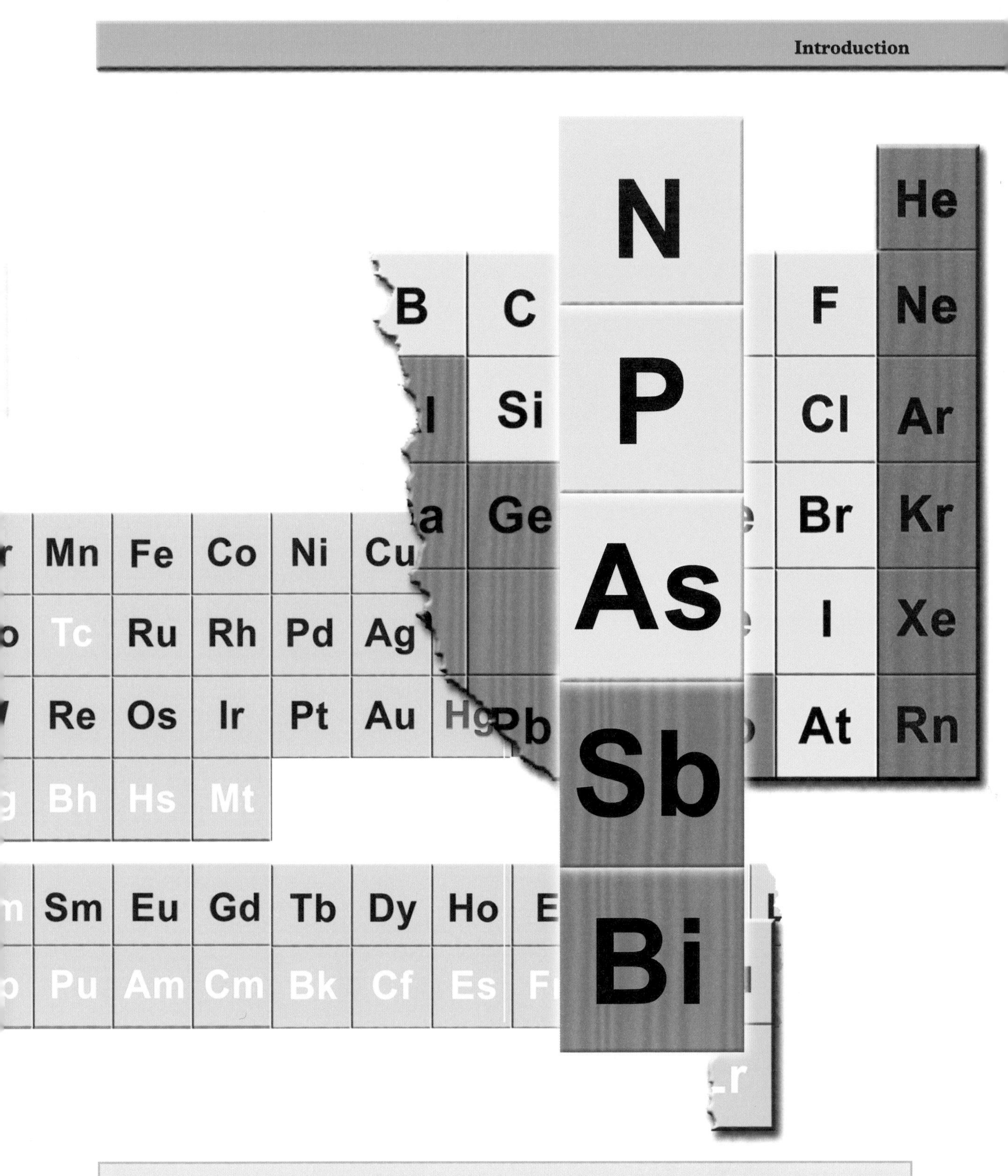

The periodic table makes finding information about the elements easy for students and scientists. The nitrogen group is found on the right side of the periodic table.

dangerous and yet so vital? The other elements of the nitrogen group also have interesting stories. Which one is highly poisonous but is sometimes used to treat cancer? Which one is an ingredient in the world's strongest acid? Read on to find out.

Chapter One
Meet the Nitrogen Family

Our atmosphere is made up of about 78 percent nitrogen by volume and 76 percent by weight. Nitrogen is considered an inert element because nitrogen gas normally does not interact with other substances. Individual atoms of nitrogen are highly reactive. And because of this, they are never found alone. In pure nitrogen, two atoms link together, which makes them highly stable. These atom pairs are called diatomic molecules. Nitrogen in the air is made up of these diatomic molecules.

Most scientists generally agree that Scottish chemist Daniel Rutherford discovered nitrogen in 1772. After Rutherford used previously identified processes to remove oxygen and carbon

Daniel Rutherford first isolated nitrogen while he was a student at the University of Edinburgh.

dioxide from the air in a sealed container, he noticed that there was still a gas in the container. Further experimentation showed that the gas could not support life or combustion. This gas was nitrogen.

Phosphorus

Phosphorus is a highly reactive solid, and it is never found free in nature. This element is perhaps most well known for its ability to glow in the dark. For many years, the main use for phosphorus has been in the production of fertilizer. Originally, phosphorus collected from animal bones was used to fertilize crops. A substance called guano (bat and bird droppings) is rich in phosphorus. It has been used as a fertilizer for hundreds of years. Today, most fertilizers are made from the phosphorus refined from phosphates—minerals that contain phosphorus.

Around 1669, German alchemist Hennig Brand was searching for the "philosopher's stone," a mythical substance that many alchemists believed could turn inexpensive metals into gold. To discover phosphorus, Brand collected more than sixty buckets of urine and allowed it to putrefy. He then boiled and filtered the urine until it formed a paste. Next, he heated

What's in a Name?

French chemist Jean-Antoine Chaptal gave nitrogen its name in 1790. He discovered that the element was present in certain known substances called nitrates, particularly the compound potassium nitrate (KNO_3), which was one of the first substances known to contain nitrogen. The common name for potassium nitrate is niter (also called saltpeter). The word "nitrogen" comes from the French words for "niter producing."

the substance and drew the vapors into water in the hope of making gold. Instead, he produced a white, waxy substance that glowed in the dark and burst into flames when exposed to oxygen. Years later, the substance was named phosphorus, which comes from the Greek words meaning "light-bringing."

Arsenic

Arsenic is usually a shiny, brittle, silver gray solid. It tarnishes to a dull black when exposed to oxygen. It may also look yellow or black depending on what kind it is. Pure arsenic can be found in very minute amounts in the earth's crust, but it is usually found combined with other solids. When arsenic is heated to 1,141 degrees Fahrenheit (616 degrees Celsius), it changes directly from a solid to a gas. This action is called sublimation. When under high pressure, however, arsenic can be forced to melt at 1,503°F (817°C).

Arsenic, which was one of the earliest discovered elements, has a long and notorious history as a poison. As far back as the fourth century BCE, arsenic was used—mainly by wealthy Roman noblemen and women—as a way to get rid of enemies. It was also a

In 1840, Marie Lafarge was convicted of using arsenic to murder her husband, Charles. She was one of the first people to go to prison for arsenic poisoning.

popular method of murder chosen by wealthy women of the Renaissance period. At the time, it was easy to obtain because it was a common form of rat poison, as well as a popular cosmetic. Most scientists believe pure arsenic was first isolated in 1250 by German theologian, philosopher, and scientist Albertus Magnus. Despite being a dangerous element, arsenic has been used for a wide variety of products, from dyes to electronics.

Antimony

The most stable form of antimony looks very similar to arsenic and is usually found in compounds with the same elements that arsenic is found

Many Muslim men apply the cosmetic kohl on the first day of Ramadan, a holy month that includes fasting and prayer. Muslims abstain from eating and drinking from dawn until dusk during Ramadan, but the use of kohl is usually allowed.

with. It is hard to tell them apart without special tests. The mineral stibarsen is a mixture of arsenic and antimony. It often takes a trained mineralogist to tell the difference between arsenic, antimony, and stibarsen.

Scientists are not sure who first discovered antimony, although antimony compounds were known to people as far back as 3000 BCE, and perhaps earlier. Many ancient women, particularly those of Egypt, used antimony sulfide as a cosmetic to darken the areas around their eyes. In this use, it is often referred to as kohl.

Throughout the Middle Ages, antimony was a popular material among alchemists. Writings from the 1500s describe how to isolate the element from its compounds. The French chemist Nicolas Lémery conducted the earliest studies of the element in the early 1600s. Swedish chemist Jons Jakob Berzelius gave antimony its chemical symbol: Sb. Berzelius used an abbreviation of the Latin word *stibium*, which comes from *stibnite*, the Greek word for antimony sulfide.

Boom!

Nitrogen is able to form many compounds when N_2 molecules are separated into atoms, both naturally and synthetically. Some nitrogen compounds break down into individual atoms when they burn, releasing a large amount of nitrogen atoms. These atoms come together quickly to form N_2 molecules. The formation of the triple bond between nitrogen atoms releases an immense amount of energy. The result is an explosion. Explosive nitrogen compounds are used to make bombs for peaceful and violent purposes. They are also used to make rocket fuel.

Bismuth

Bismuth is a heavy, brittle metal. It is grayish white with a pinkish hue. Bismuth crystals are popular among mineral collectors because they are beautiful and can be made easily in a lab or even a kitchen. They are well known for their regular angles and an oxide tarnish that appears iridescent in light. Although bismuth is a metal, it is sometimes regarded as a "poor metal" because it is brittle and because it is a very poor conductor of heat and electricity.

Bismuth was probably known of in ancient times, although it was usually confused with tin (Sn), lead (Pb), zinc (Zn), and antimony up

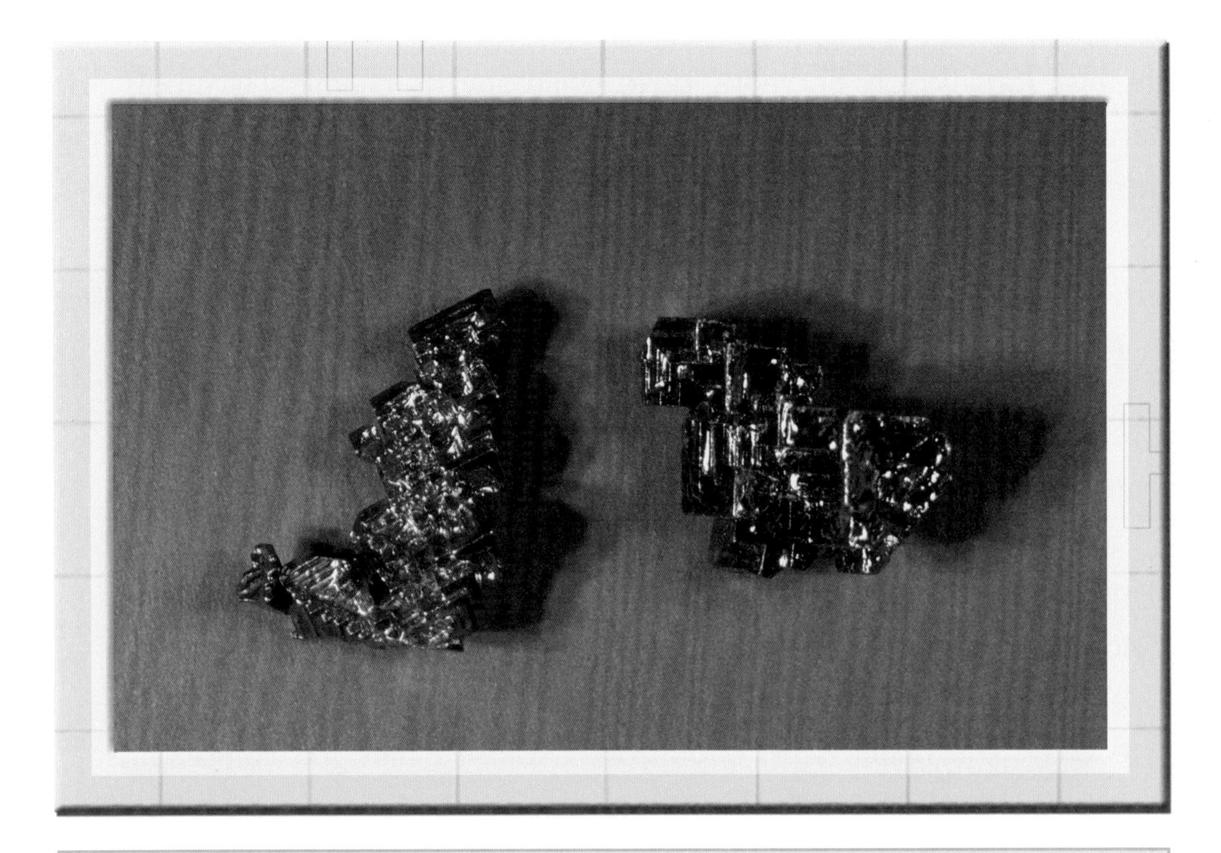

Pure bismuth looks very different from other metals. Some kinds of fishing sinkers are made of bismuth, and they can be melted down to produce beautiful crystals like these.

until the eighteenth century. Alchemists experimented with it and sometimes separated it from bismuth compounds. Around 1450, bismuth alloys were used to make type for printing presses. It is perfect for this application because it expands when it changes from a liquid to a solid, and it is able to completely fill the molds used to make fine type.

In 1753, French chemist Claude François Geoffroy became the first scientist to isolate pure bismuth. He is often credited as the discoverer of bismuth. The origin of the word "bismuth" is uncertain, although some scientists believe it could have come from the German words *weisse masse*, meaning "white material."

Chapter Two
Inside the Group

Atoms are extremely small pieces of matter that make up everything in the universe. Millions of atoms can fit into a space the size of the period at the end of this sentence. Atoms are made of smaller pieces of matter called subatomic particles. At the center of every atom is the nucleus—a cluster of subatomic particles called protons and neutrons. Most of an atom's weight is localized in the nucleus.

Electrons, which are much smaller and lighter than protons and neutrons, are subatomic particles found in the space around the nucleus. Each electron has a negative electrical charge. They are attracted to the protons in the nucleus, each of which has a positive electrical charge. Neutrons do not have an electrical charge. Most of the space inside an atom is occupied by the electrons.

Isotopes and Ions

An atom of a specific element always has the same number of protons. Nitrogen, for example, always has seven protons. The number of protons in an atom is its atomic number. An atom may have a different number of neutrons or electrons. Atoms with different numbers of neutrons are called isotopes. Many isotopes are radioactive and are often referred to as radioisotopes. If you add the number of protons and neutrons in the

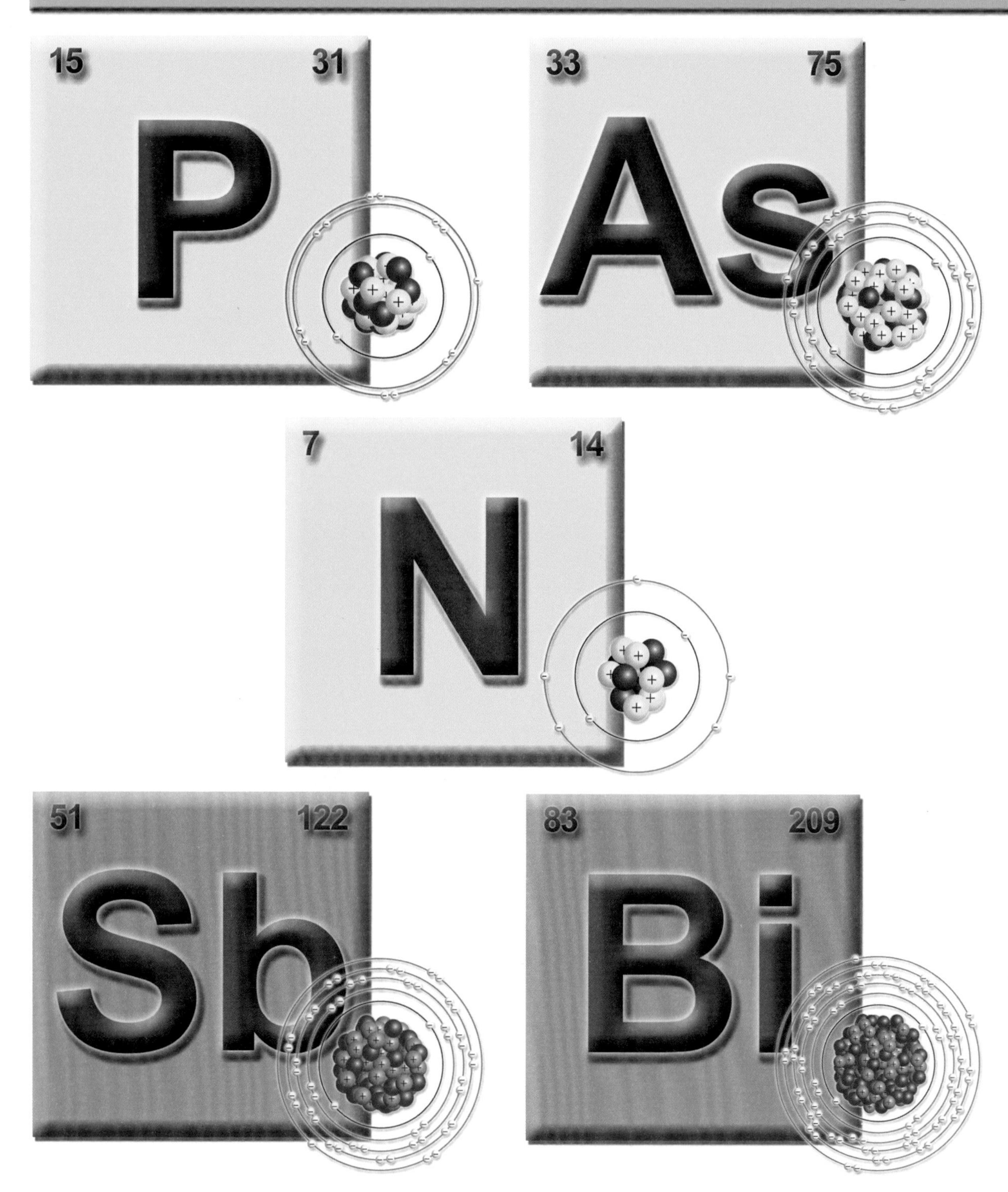

Each of the elements on the periodic table is represented by a square containing basic information—chemical symbol, atomic number (upper-left corner), and atomic weight (upper-right corner).

nucleus of an atom, you have that atom's atomic weight. The atomic weight of a nitrogen atom with seven protons and six neutrons is 13. The atomic weight shown for each element in the periodic table is an average weight for all of the element's isotopes.

An atom that gains or loses one or more electrons is called an ion. Atoms normally have the same number of electrons and protons, which makes them electrically neutral. Atoms that have extra electrons have a negative charge and are called anions. Atoms with fewer electrons than protons have a positive charge and are called cations.

Atomic Bonds

Atoms are able to form different substances by linking together in different combinations. The links between atoms are called chemical bonds. Atoms form chemical bonds with their outermost electrons. Electrons form

Bismuth 209

Radioisotopes are isotopes that give off energy in the form of subatomic particles. Radioisotopes decay to form other isotopes, some of which are also radioactive and others that are stable. The rate of decay is expressed with a unit called a half-life. This is the amount of time it takes for a radioisotope to lose half of its radioactivity.

Scientists once believed that bismuth 209 (Bi-209) was a stable isotope of bismuth. However, they later discovered that Bi-209 has a half-life of 1.9 x 1019 years. That is very much longer than the age of the earth and even of the universe itself. All of the bismuth found in nature is Bi-209.

layers around the nucleus that scientists call shells. Each shell can hold a limited number of electrons. When an inner shell is full, more electrons gather in the next shell. Atoms with full outer shells are less likely to form bonds with other atoms. Atoms that have an outer shell that is not full often easily form bonds with other atoms.

Atomic Structure and Physical Characteristics

The way an element acts depends on the number of protons its atoms contain. Even a difference of just a few protons can result in very noticeable

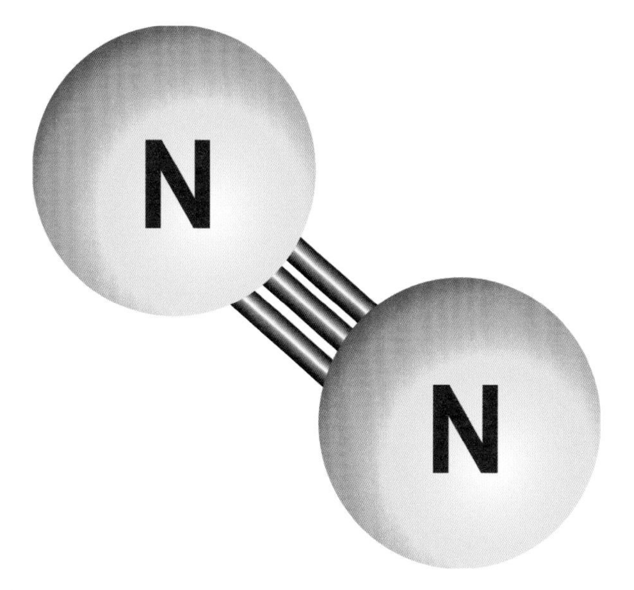

Diatomic nitrogen molecules are formed by very strong triple bonds. All of the nitrogen in Earth's atmosphere is in the form of diatomic particles. It is very difficult to separate the atoms in an N_2 molecule.

differences in the way two elements act. For example, nitrogen—whose atoms contain seven protons—is a colorless, odorless gas at room temperature. Carbon, on the other hand, is usually a black solid at room temperature, even though its atoms have just one less proton.

A single nitrogen atom has five electrons in its outer shell. The outermost shell can hold eight electrons. Because the outermost shell is not full, nitrogen atoms are highly reactive and form compounds with many other elements. Because the individual atoms are very reactive, pure nitrogen is found in a diatomic form. Two reactive atoms combine into a nonreactive molecule. The two atoms in a nitrogen molecule (N_2) form three chemical bonds between each other, making the molecule very stable and difficult to separate into individual atoms.

Group 15

All group 15 atoms have five electrons in their outer shells and can form numerous compounds with other elements. These compounds are often very stable because the group 15 elements form multiple bonds with other elements. Physically, however, the group 15 elements are very different. The further down the group, the more metallic the elements become.

Allotropes

Some elements exist in two or more different forms. These different forms of the same element are called allotropes. They differ in the way the atoms are arranged. Phosphorus has three main allotropes and several other less common allotropes. White phosphorus is made up of molecules containing four atoms and six bonds. It is a white, waxy substance that turns yellow and can spontaneously ignite when exposed to air. It

Group 15 at a Glance

	Nitrogen	Phosphorus	Arsenic	Antimony	Bismuth
Chemical symbol	N	P	Ar	Sb	Bi
Atomic number	7	15	33	51	83
Atomic weight	14	31	75	122	209
Protons	7	15	33	51	83
Electrons	7	15	33	51	83
Neutrons	6	16	42	71	126
Melting point	–346°F (–210°C)	111°F (44°C)	1,503°F (817°C)	1,167°F (631°C)	521°F (272°C)
Boiling point	–321°F (–196°C)	536°F (280°C)	1,135°F (613°C)	2,889°F (1,587°C)	2,840°F (1,560°C)

produces a greenish glow when exposed to oxygen. White phosphorus is highly toxic.

Red phosphorus forms when white phosphorus is exposed to sunlight or heated above 482°F (250°C). This causes the atoms to bond in a much more disordered arrangement, which gives the substance much greater stability. Red phosphorus is a red, powdery substance that

When white phosphorus comes into contact with oxygen, a chemical reaction takes place that creates sparks, bright light, and white smoke.

does not catch fire as easily as white phosphorus. Other allotropes include violet and black phosphorus.

Nitrogen has several unstable allotropes. Arsenic has at least three allotropes. Antimony, which is similar to arsenic in several ways, has four. Bismuth may have several.

Match Point

White phosphorus and other chemicals were used to make friction matches. Rubbing the matches on a rough surface made enough heat to ignite the chemicals. Red phosphorus, discovered in 1845, soon replaced white phosphorus because it is less volatile and poisonous. Today's "safety matches" have just a small amount of red phosphorus on the tips and must be rubbed on a specially prepared surface to light. It is interesting to note that many safety matches also contain antimony trisulfide (Sb_2S_3), which burns and helps ignite the wooden match.

Metals, Nonmetals, and Metalloids

The periodic table groups similar types of elements near each other. Metals, which take up about three-quarters of the table, are usually shiny and solid at room temperature (mercury [Hg] is the only metal that is liquid at room temperature). Metals are also good conductors of heat and electricity because their atoms easily exchange electrons, which moves electrical charges from one atom to another. The elements on the right side of the table are nonmetals, some of which are gases at room temperature. They are usually poor conductors of heat and electricity.

The elements of the nitrogen family can look very different from one another. From left to right in the top row are samples of antimony, white phosphorus, and arsenic. From left to right on the bottom are samples of red phosphorus and bismuth.

Between the metals and nonmetals is a narrow, diagonal band of elements called metalloids or semimetals. They exhibit some characteristics of both metals and nonmetals. Metalloids usually make good semiconductors, and they also usually have several allotropes.

Group 15 contains at least one of each of these kinds of elements. Nitrogen (which is a gas at room temperature) and phosphorus are nonmetals. Bismuth is a metal. Arsenic and antimony are metalloids.

Chapter Three
Refining the Elements

We use the earth's many elements for a wide variety of purposes. However, elements aren't always available in pure form in nature. Often, we need to extract an element from ores found in the earth's crust. Others are taken from the atmosphere or from seawater. The process used to separate the group 15 elements from their natural sources is called refining.

Distilling Nitrogen

The air is our greatest source of nitrogen, but nitrogen needs to be separated from the other elements in the air to make it pure. Pure nitrogen is produced using a process called distillation. Although nitrogen is normally a gas, it can be liquefied very easily. To make liquid nitrogen, the element needs to be cooled to less than −321°F (−196°C). When liquid nitrogen rises above this temperature it boils, or turns back into a gas.

To obtain liquid nitrogen, manufacturers liquefy large quantities of air in a glass column separated by permeable partitions. This container is called a fractional column. Air is made mostly of nitrogen and oxygen gas, but it also contains significant amounts of argon (Ar). Oxygen liquifies at −297°F (−183°C), and argon liquifies at −303°F (−186°C).

Chef Pino Maffeo uses liquid nitrogen to make a special dessert. He dips spoons of yogurt into the nitrogen to form a frozen outer shell, and then he injects jelly into the center.

Once the air is completely liquefied at −321°F (−196°C), it is allowed to warm up again very slowly. The nitrogen boils first and the gas rises to the top of the glass container, where it is collected and removed as the liquid oxygen and liquid argon drip into lower sections of the column.

To make distillation easier to accomplish, the air is also cooled under greater-than-normal pressure, which raises the temperature at which it liquifies. Liquid nitrogen is also stored under pressure to help keep it in liquid form.

Phosphorus

Since phosphorus is never found pure in nature, it must be refined from the minerals that contain it. Phosphate rocks, the most common source of phosphorus, usually contain the mineral apatite ($Ca_5F(PO_4)_3$). One form of apatite is found in animal bones, which were at one time used as a source of phosphorus.

Thankfully, people who produce phosphorus no longer use the archaic process described in chapter 1. Today, most pure phosphorus is obtained from phosphate ores. Phosphate ore is mixed with silica (which is sand)

Apatite, shown here, forms as long, six-sided crystals. The crystals are usually greenish brown with white streaks.

and carbon in an electric furnace. An electric current is passed through the mixture until it melts. Several gases are produced, including pure phosphorus. The phosphorus gas condenses in the form of white phosphorus.

Arsenic

Pure arsenic is very rare in nature. It is most often found in compounds containing other elements, particularly antimony, sulfur, iron, nickel (Ni), and silver (Ag). The main ore of arsenic, arsenopyrite—or iron arsenic sulfide (FeAsS)—contains about 46 percent arsenic.

Commercial arsenic is usually produced as a byproduct when smelting other metals, especially iron, copper (Cu), and lead. It is often obtained from iron arsenide ($FeAs_2$) or arsenopyrite. These ores are heated to about 700°F (370°C) in the absence of air. During this process, arsenic in the ore sublimes, or changes directly to a gas. It quickly condenses and is collected. Arsenic compounds also form slag—liquid impurities—on the top of the molten iron. The slag is skimmed off and further refined to produce pure arsenic. When arsenic is heated in the presence of air, it combines with oxygen to form a poisonous white gas called arsenic trioxide.

This is an aerial view of a geothermal pool in New Zealand called Champagne Pool. The orange deposits around the pool get their color from high concentrations of the minerals stibnite (Sb_2S_3) and orpiment (As_2S_3).

Antimony

Stibnite is the predominant antimony ore from which the element is commercially refined. Small amounts of antimony are often separated from other minerals when they are smelted to produce pure metals. These metals include silver, gold, copper, and lead ores. Antimony is often attained at the same time arsenic is refined because they often appear in the same ores.

China is the largest producer of antimony in the world. There are no currently operating antimony mines in the United States, although the element is collected during the smelting of copper and silver. Much of the antimony produced by the United States comes from the recycling of electrical storage batteries, which often contain antimony.

Freezing with Liquid Nitrogen

Manufacturers liquefy nitrogen to purify it, but they also sell nitrogen in its liquid form. You may have seen a scientist experimenting with liquid nitrogen. It is so cold that a rose dipped in it freezes instantly and will shatter when tapped against a hard surface. Liquid nitrogen boils quickly when exposed to room temperature. Cold nitrogen gas coming from the liquid causes water vapor in the air to condense into fog.

Liquid nitrogen is most useful when prolonged and deep cooling is necessary, as in the transportation of food products, blood, and medical supplies over long distances. It is also used in a branch of science called cryogenics, which is the study of the production of very low temperatures and the way materials act at those temperatures.

Bismuth

Bismuth is usually attained as a byproduct in the mining and refining of other, more abundant materials, particularly lead, but also copper, tin, silver, gold, and tungsten (W). Bismuthinite (Bi_2S_3) and bismite (Bi_2O_3) are the most important bismuth ores.

Pure bismuth is usually refined from lead using the Betterton-Kroll process, or Betts process. First, calcium and magnesium (Mg) are added to molten lead-bismuth. The calcium and magnesium form bismuth compounds that have higher melting points and lower densities than the lead. As a result, they form a dross, which is solid impurities floating on the top of a molten material. The dross is skimmed off the pure molten lead. Next, the compounds are treated with chlorine (Cl). The chlorine forms compounds with the calcium and magnesium, freeing up the bismuth.

Chapter Four The Compounds

A compound is a combination of two or more elements. The atoms of the elements are chemically bonded, meaning that their atoms link together to form larger pieces of matter called molecules.

Compounds seldom exhibit characteristics of the elements of which they are comprised. Usually, compounds take on new characteristics that the original elements did not have at first. Compounds are frequently more helpful to us than pure elements. On the other hand, many compounds can harm people.

Nitrogen Compounds

Nitrogen compounds are used to make explosives, jet fuel, cleansers, fertilizers, and many other products. The most important nitrogen compound is ammonia (NH_3). Approximately 40 percent of the world's population depends on crops grown using fertilizers that contain ammonia. Ammonia is also used in the manufacture of plastics, fabrics, pharmaceuticals, refrigerants, metals, and cleansers. Pure ammonia is a gas at room temperature. It has a very strong odor, and it will cause eyes to water. If more than a tiny amount is inhaled, it will

This diagram demonstrates the Haber-Bosch process used to make ammonia. Nitrogen (blue) and hydrogen (red) are combined to produce the chemical (yellow).

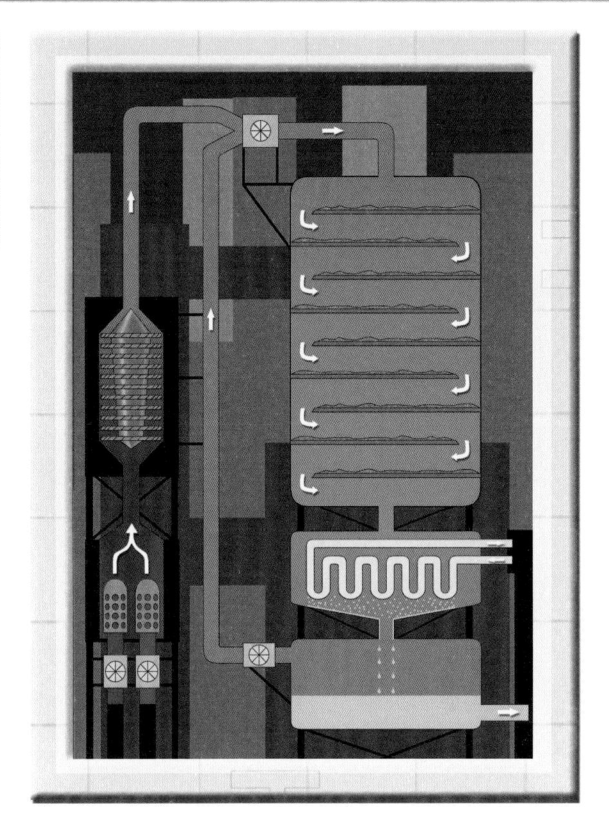

cause difficulty in breathing. Ammonia is often used commercially in a very cold liquid form. Liquid ammonia is very caustic, and people working with ammonia should wear protective gear to avoid injuries.

Ammonia is found in small traces in nature. It is found in the air as a result of the decomposition of dead plants and animals. It is also found in animal waste. Many plants use ammonia in the soil to synthesize amino acids, which are vital in the production of protein. At one time, ammonia was made by boiling animal horns and hoofs.

Commercial ammonia is made via the Haber-Bosch process, a method developed by German scientists during World War I. Nitrogen and hydrogen (H) gases are combined under pressure at a temperature between 842°F and 1,112°F (450°C and 600°C), using iron as a catalyst. The ammonia gas becomes liquid after rapid cooling. The higher the pressure, the more pure the ammonia, but the longer the process takes. Not all of the nitrogen and hydrogen gases are used up during this process. Remaining gases are recycled to make more ammonia.

This laboratory machine is used to test the toxic effects of organophosphate chemicals on living tissues. Organophosphate compounds are common chemicals that contain phosphorus and carbon, among other elements.

Phosphorus Compounds

The most important phosphorus compound is phosphoric acid (H_3PO_4). Its main use is in the manufacture of fertilizers. It is also used to remove rust from steel and as a polish for some metals, such as aluminum. It is added to cola beverages to enhance the tangy flavor. Other uses include the manufacture of soaps, refining sugar, water treatment, and animal feeds, to name just a few. Other important phosphorus compounds include phosphorus pentasulfide (P_2S_5) and phosphorus trichloride (PCl_3), both of which are used in the manufacture of pesticides.

What Is NO_x?

The term "NO_X" refers to the group of gases that contain nitrogen and oxygen in varying amounts. These gases include nitrous oxide (N_2O), nitric oxide (NO), and nitrogen dioxide (NO_2). Nitrogen and oxygen don't normally react with each other, but they form compounds during combustion at high temperatures, as happens, for example, in automobile engines. Like other automobile emissions, it is a poison that contributes greatly to air pollution and breathing problems in many people. In urban areas, nitrogen dioxide contributes to a layer of reddish brown gas commonly called smog. NO_X gases are also categorized as greenhouse gases, which contribute to global warming.

Many phosphorus compounds are toxic, and their use has waned over the years. For example, a group of phosphorus compounds called organophosphates has been used as insecticides. They are made by combining phosphorus with compounds that contain carbon. While insecticides made with organophosphates are still used, many people are concerned that they are too dangerous to use around food supplies. Other organophosphates were developed as nerve gases prior to World War II. However, they were never used in warfare.

Arsenic Compounds

Arsenic compounds are used for numerous industrial purposes. They are used to purify other metals and make them harder. They are also used to remove unwanted colors from glass during glass production.

All arsenic compounds are poisonous and must be handled with care. In the past, arsenic compounds were widely used as wood preservatives for deck, porches, and even playground equipment. However, they were suspected to cause poisoning in people who touched the wood. They have also been found to contaminate soil that comes into contact with the wood. Today, arsenic-treated wood is used for only very few applications, such as roofing material.

Superfast Electronics

Arsenic is used in the manufacture of microchips and semiconductors. A semiconductor is a substance that can act both as an electrical conductor

Arsenic compounds are used to make photodiodes, such as the one shown here. A photodiode converts light to electrical current.

and an electrical insulator, depending on the direction that electricity is applied to it. It accomplishes this unusual feat by the way it is made, with ingredients called doping agents.

Arsenic is sometimes used as a doping agent in microchips. The compound gallium arsenide (GaAs) has become an important semiconductor used in some of the fastest microchips on the market. It is also used to make tiny lasers that convert electric signals into pulses of light. It is used as well in the manufacture of tiny lights called light-emitting diodes (LEDs). Gallium arsenide lasers and diodes are used in many products today, including DVD players, cell phones, computers, and solar panels, to name just a few.

Antimony Compounds

The most important commercial antimony compound is antimony trioxide (Sb_2O_3). It is produced by burning pure antimony in the air or by heating ores that contain antimony in the air. Antimony trioxide is used in the manufacture of flame retardants, opaque glass, and a pigment named antimony white. Like arsenic, other antimony compounds are increasingly being used to make semiconductors. Antimony pentoxide (Sb_2O_5) is commonly used as a fire retardant for plastics and textiles.

Fluoroantimonic Acid

Mixing equal parts hydrogen fluoride (HF) and antimony pentafluoride (SbF_5) creates the strongest known acid: flouroantimonic acid. It is much stronger than 100 percent sulfuric acid and is in a group that scientists call superacids. There are many uses for superacids, and as you might imagine, they are highly dangerous. Flouroantimonic acid is used to create and maintain carbon-based cations (atoms with a positive charge). Carbon-based cations are used in the manufacture of plastics and oil products. Using a superacid to stabilize carbon-based cations allows researchers more time to study them.

Steel is an alloy made mostly of iron and a small amount of carbon. It is very strong. Steel cable is used in the manufacture of bridges and towers.

Bismuth Alloys

An alloy is a mixture of two or more elements, at least one of which is a metal. Alloys are similar to compounds. However, the atoms in them do not form chemical bonds. The resulting mixture usually displays improved metallic characteristics. Some alloys, such as steel, are much stronger than pure metals. Others, such as stainless steel, are more resistant to corrosion and rust.

Bismuth is most often used to make fusible alloys, which are metal mixtures with very low melting points. Antimony is also used to make fusible metals. Two common bismuth alloys are Field's metal—which contains bismuth, indium (In), and tin—and Wood's metal—containing bismuth, lead, tin, and cadmium. These unique mixtures form substances that melt at low temperatures, both below the boiling point of water. Field's metal melts at 144°F (62°C). Wood's metal melts at 158°F (70°C).

Field's metal and Wood's metal are used in many fire sprinklers. Small plugs made out of these metals are used to keep water from coming out of the sprinklers. However, when the plugs heat up due to fire, they quickly melt. This creates a shower of water to put out the flames. Fusible bismuth alloys are also used as solder, which is a substance that can be easily melted to join, or fuse, two metal surfaces.

Chapter Five
The Elements in Our World

Some elements are vital for human health. Some, called trace elements, are needed in very small quantities. Others are often found harmlessly in the body but play no role in our health. Still others are dangerous to our health, even in very small amounts. Following are the roles the group 15 elements may or may not play in the human body.

Nitrogen

All living things need nitrogen to survive. It is an important ingredient in amino acids, which are the building blocks of protein. Humans need protein to build and repair cells, particularly those in the muscles, skin, and bones.

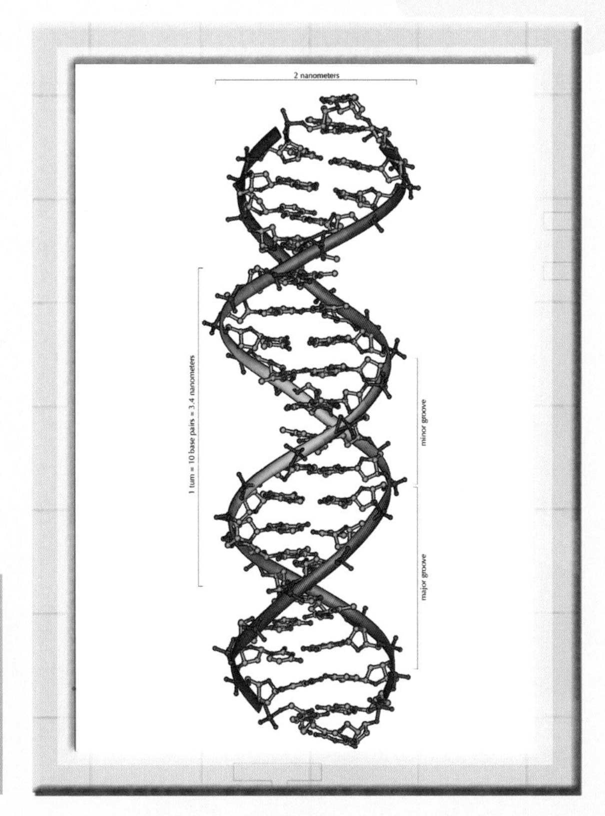

Shown here is an artist's representation of DNA, or deoxyribonucleic acid. DNA helps determine personal traits, such as eye color, hair color, and height.

Nitrogen is also an important component of DNA and RNA. You might think of DNA and RNA as instruction manuals for our bodies. However, animals and most plants are not able to take nitrogen from the air. We would not be able to get the nitrogen we need to live if it weren't for the nitrogen cycle.

Phosphorus

Like nitrogen, phosphorus is a component of DNA and RNA and an essential element for all living cells. Phosphates are compounds found in all living things. The phosphate ion (PO_{43}–) is particularly abundant in living things, and it is easy for the body to make and break apart. Most of the phosphate we get from the foods we eat combines with calcium in the bones. It is also found in the blood.

Too much phosphorus in the body can result in health problems, particularly kidney damage and osteoporosis, a weakening of the bones. Pure phosphorus is highly poisonous. In the 1800s and early 1900s, workers in factories who came into contact with white phosphorus vapors developed a serious illness called phossy jaw. This is a decaying of the bones in the jaw. Most people who contracted phossy jaw had to have their jaws removed, and they ultimately died from complications.

Arsenic

All arsenic compounds are poisonous: just 100 milligrams of arsenic is enough to kill a human being. Arsenic should never be touched, ingested, or inhaled. It inhibits the body's metabolizing processes, which are a series of chemical reactions that support growth and life. There are many symptoms of arsenic poisoning, including stomach pains, vomiting,

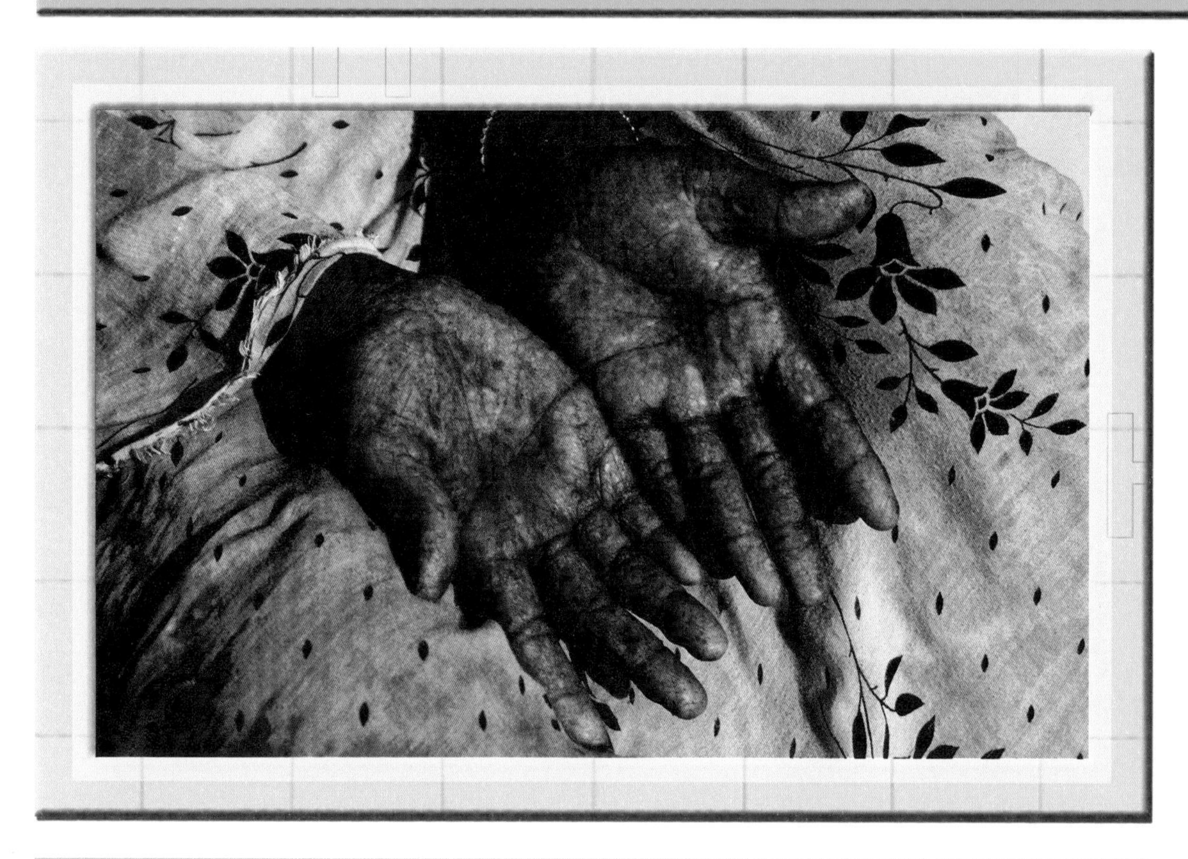

In Bangladesh, a woman displays the effects of drinking arsenic-tainted water. This has become a common sight in that area of the world. Many countries are working to help the people of Bangladesh find new, clean sources of water.

diarrhea, internal bleeding, shivering, headaches, confusion, coma, heart failure, and death.

In some areas of the world—particularly Bangladesh—arsenic-tainted drinking water has become a serious problem. The symptoms of chronic poisoning include white lines spanning the fingernails and toenails, discoloration of the skin, warts and bumps, and gangrene. These symptoms can appear as early as two weeks or may not appear for years. Long-term symptoms can include cancer, heart disease, and diabetes. Very few drugs contain arsenic. The drug Trisenox, which contains arsenic trioxide, is sometimes used to treat leukemia.

The Nitrogen Cycle

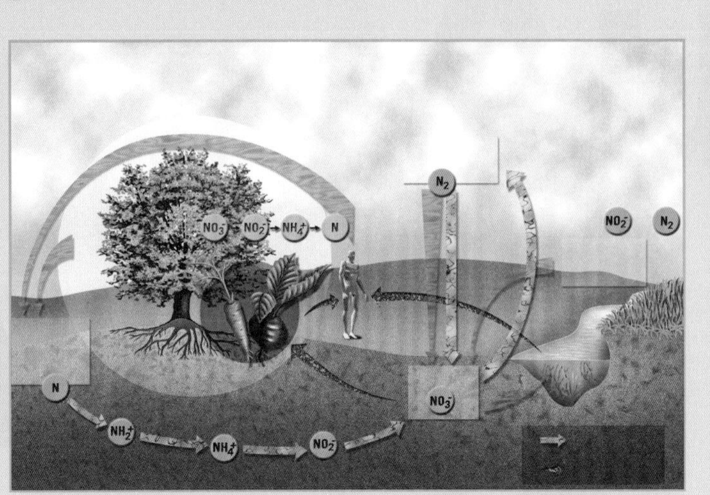

This diagram demonstrates the ways in which nitrogen is cycled through the environment.

Living things are not able to take pure nitrogen from the air. Energy is needed in order to convert the pure nitrogen into nitrogen compounds. This process, which is called nitrogen fixing, can happen in two ways. First, lightning heats up the air and allows nitrogen and oxygen to form compounds, which reach the soil in rain. Second, and more commonly, some kinds of bacteria, fungi, and algae take nitrogen from the air and convert it into nitrogen compounds that plants can use. One of the most common nitrogen-fixing bacteria is called *Rhyzobium*, which lives in nodules on the roots of plants in the legume family (peas, peanuts, clover, etc.). Much of the compounds that *Rhyzobium* makes (particularly ammonia) is left in the soil, and other plants can get it there. Animals and people get the nitrogen compounds they need from the plants and animals they eat.

When living creatures die, their tissues are broken down by the natural process of decomposition. This process returns many elements, including nitrogen, back into the environment. Nitrogen is also returned to the environment in animal waste.

Antimony

Antimony poisoning is similar to arsenic poisoning, but it is not as severe or common. The most common occurrence of antimony poisoning is when people who work with it inhale its vapors. It can also enter the body by way of contaminated food and water. Small amounts of antimony can cause eye, skin, and lung irritation. Prolonged exposure can cause diarrhea, severe vomiting, heart problems, and lung disease. Antimony trioxide is used in the manufacture of some plastic bottles. Some scientists are concerned that the antimony may contaminate drinking water and the soils around garbage dumps.

Bismuth

Bismuth is usually not harmful to people in small doses. Continuous exposure can cause minor kidney damage, and large doses can be fatal. Bismuth alloys are used in many kinds of plumbing solder, so plumbers may breathe in bismuth fumes when working with it.

Bismuth subsalicylate is commonly used in over-the-counter medications for upset stomach. Children with chicken pox should not take medicines that contain this compound, as it has been linked to a rare disease called Reye's syndrome, which can be fatal. Some brands of lipstick contain bismuth oxychloride to create a shiny, metallic color. Some people may experience an allergic reaction to this compound.

The Periodic Table of Elements

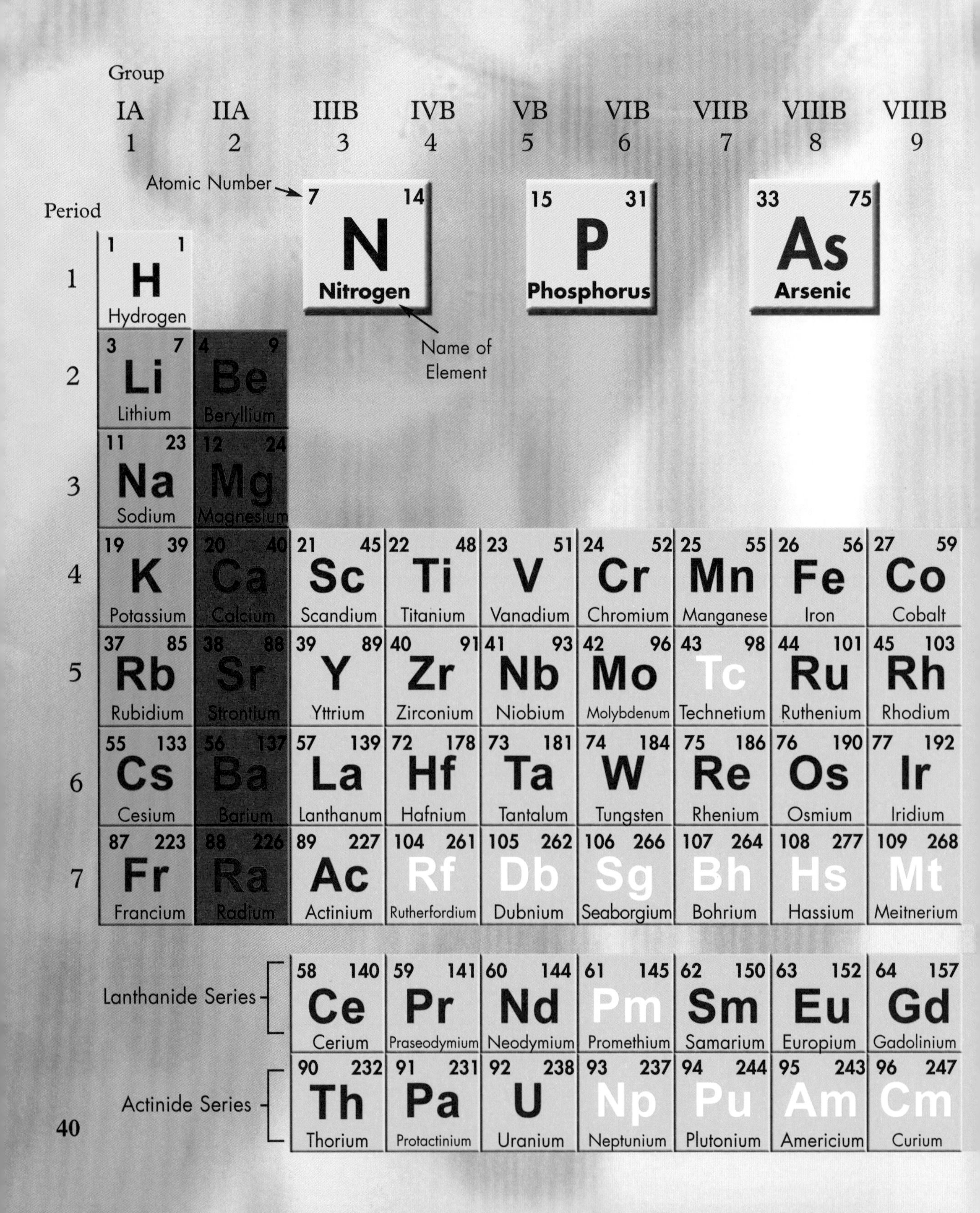

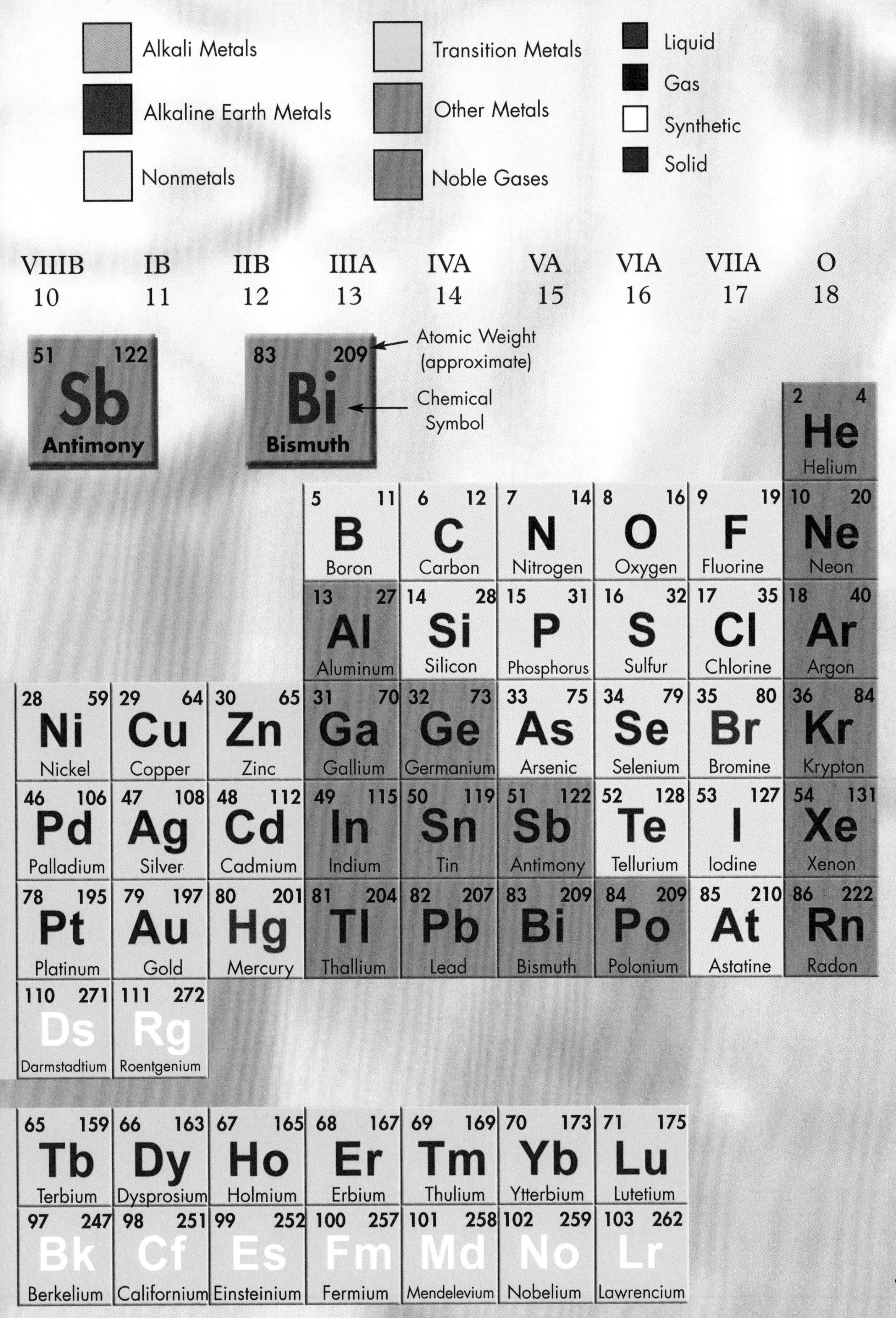
Alkali Metals
Alkaline Earth Metals
Nonmetals
Transition Metals
Other Metals
Noble Gases
Liquid
Gas
Synthetic
Solid
VIIIB 10
IB 11
IIB 12
IIIA 13
IVA 14
VA 15
VIA 16
VIIA 17
O 18
51 122 Sb Antimony
83 209 Bi Bismuth
Atomic Weight (approximate)
Chemical Symbol
2 4 He Helium
5 11 B Boron
6 12 C Carbon
7 14 N Nitrogen
8 16 O Oxygen
9 19 F Fluorine
10 20 Ne Neon
13 27 Al Aluminum
14 28 Si Silicon
15 31 P Phosphorus
16 32 S Sulfur
17 35 Cl Chlorine
18 40 Ar Argon
28 59 Ni Nickel
29 64 Cu Copper
30 65 Zn Zinc
31 70 Ga Gallium
32 73 Ge Germanium
33 75 As Arsenic
34 79 Se Selenium
35 80 Br Bromine
36 84 Kr Krypton
46 106 Pd Palladium
47 108 Ag Silver
48 112 Cd Cadmium
49 115 In Indium
50 119 Sn Tin
51 122 Sb Antimony
52 128 Te Tellurium
53 127 I Iodine
54 131 Xe Xenon
78 195 Pt Platinum
79 197 Au Gold
80 201 Hg Mercury
81 204 Tl Thallium
82 207 Pb Lead
83 209 Bi Bismuth
84 209 Po Polonium
85 210 At Astatine
86 222 Rn Radon
110 271 Ds Darmstadtium
111 272 Rg Roentgenium
65 159 Tb Terbium
66 163 Dy Dysprosium
67 165 Ho Holmium
68 167 Er Erbium
69 169 Tm Thulium
70 173 Yb Ytterbium
71 175 Lu Lutetium
97 247 Bk Berkelium
98 251 Cf Californium
99 252 Es Einsteinium
100 257 Fm Fermium
101 258 Md Mendelevium
102 259 No Nobelium
103 262 Lr Lawrencium

Glossary

alchemy An early, unscientific form of chemistry that sought impossible ends, such as a way to change cheap metals into gold.

catalyst A chemical that speeds up a chemical reaction without itself undergoing any change.

caustic Burning by chemical action.

combustion The burning of a fuel.

contaminate To make something impure, dirty, or polluted.

doping agent An impurity purposely added to a semiconductor to change its electrical characteristics.

gangrene The death of tissues resulting from a lack of blood flow.

greenhouse gas A gas that contributes to the warming of the earth's atmosphere.

half-life The time it takes a radioactive substance to lose half of its radioactivity through decay.

inert Not easily changed by chemical reaction.

nodule A small lump or swelling.

permeable Allowing liquids or gases to pass through.

pesticide A substance that kills pests, such as insects.

propellant A substance that is used to move something forward, such as a rocket.

putrefy To decay with a foul smell.

radioactive Referring to a substance that emits energy in the form of a stream of subatomic particles and energy as it decays.

semiconductor A solid material that has the electrical conductivity between that of a conductor and an insulator.

smelt To melt ore and separate metal from it.

For More Information

American Chemical Society
1155 Sixteenth Street NW
Washington, DC 20036
(800) 227-5558 (U.S. only)
(202) 872-4600 (outside the U.S.)
Web site: http://www.chemistry.org
This organization provides news, information, and resources for chemists. Members receive extra benefits, such as career services, networking opportunities, and free subscriptions to industry magazines.

American Chemistry Council (ACC)
1300 Wilson Boulevard
Arlington, VA 22209
(703) 741-5000
Web site: http://www.americanchemistry.com
The ACC represents the leading chemical manufacturers in the country and is an advocate for the use of chemicals to improve life on the earth.

Chemical Institute of Canada
130 Slater Street, Suite 550
Ottawa, ON K1P 6E2
Canada
(888) 542-2242
Web site: http://www.cheminst.ca
The Chemical Institute of Canada is a chemistry organization that works to promote chemistry education and the use of chemistry to improve society.

International Union of Pure and Applied Chemistry (IUPAC)
IUPAC Secretariat
104 T. W. Alexander Drive, Building 19
Research Triangle Park, NC 27709
Web site: http://www.iupac.org
An international and nongovernmental advocate of the chemical sciences, the IUPAC is the recognized authority for the naming of the elements and their compounds.

U.S. Geological Survey (USGS)
12201 Sunrise Valley Drive
Reston, VA 20192
(888) ASK-USGS (275-8747)
Web site: http://www.usgs.gov
The USGS is a scientific agency of the U.S. government that studies the earth's landscapes and natural resources, as well as natural hazards that put the earth at risk.

Web Sites

Due to the changing nature of Internet links, Rosen Publishing has developed an online list of Web sites related to the subject of this book. This site is updated regularly. Please use this link to access the list:

http://www.rosenlinks.com/uept/tne

For Further Reading

Basher, Simon, and Adrian Dingle. *The Periodic Table.* London, England: Kingfisher Publications, 2008.

Manning, Phillip. *Atoms, Molecules, and Compounds.* New York, NY: Chelsea House Publications, 2007.

Miller, Ron. *The Elements.* Kirkland, WA: 21st Century, 2004.

Roza, Greg. *Arsenic.* New York, NY: Rosen Publishing Group, 2008.

Saunders, Nigel. *Nitrogen and the Elements of Group 15.* Portsmouth, NH: Heinemann, 2004.

Sommers, Michael A. *Phosphorus.* New York, NY: Rosen Publishing Group, 2007.

Tocci, Salvatore. *Nitrogen.* New York, NY: Children's Press, 2004.

Wilker, Benjamin D. *The Mystery of the Periodic Table.* Bathgate, ND: Bethlehem Books, 2003.

Bibliography

Beatty, Richard. *Phosphorus*. New York, NY: Benchmark Books, 2001.

Burros, Marion. "Chicken with Arsenic: Is That OK?" *New York Times*, April 5, 2006. Retrieved September 18, 2007 (http://www.nytimes.com/2006/04/05/dining/05well.html?ex=1301889600&en=9e2e0f56af407166&ei=5088&partner=rssnyt&emc=rss).

Cooper, Chris. *Arsenic*. New York, NY: Benchmark Books, 2007.

Global Healing Center. "The Dangers of Bismuth." April 30, 2008. Retrieved October 6, 2008 (http://www.ghchealth.com/dangers-of-bismuth.html).

Greenwood, N. N., and A. Earnshaw. *Chemistry of the Elements*. Oxford, England: Butterworth-Heinemann, 2001.

Knapp, Brian. *Nitrogen and Phosphorus*. New York, NY: Grolier Education, 1996.

Krebs, Robert E. *The History and Use of Our Earth's Chemical Elements*. Westport, CT: Greenwood Press 2006.

Lenntech. "Antimony—Sb." Retrieved October 6, 2008 (http://www.lenntech.com/Periodic-chart-elements/Sb-en.htm).

Oxlade, Chris. *Atoms*. Chicago, IL: Heinemann Library, 2002.

Index

About the Author

Greg Roza has written and edited educational materials for children for the past nine years. He has a master's degree in English from the State University of New York at Fredonia. Roza has long had an interest in scientific topics. He lives in Hamburg, New York, with his wife, Abigail, and his three children, Autumn, Lincoln, and Daisy.

Photo Credits

Cover, pp. 1, 5, 15, 17, 40–41 by Tahara Anderson; p. 7 © 2010 Jupiterimages Corporation; p. 9 Hulton Archive/Getty Images; p. 10 Khaled Fazaa/AFP/Getty Images; p. 12 © Lester V. Bergman/Corbis; p. 20 © Educational Images/Custom Medical Stock Photo; p. 21 © Charles D. Winters/Photo Researchers, Inc.; p. 24 © Brian Snyder/ Reuters/Corbis; p. 25 J & L Weber/Peter Arnold, Inc.; p. 26 © Bernhard Edmaier/Photo Researchers, Inc.; p. 29 © SPL/Photo Researchers, Inc.; p. 30 © Colin Cuthbert/Photo Researchers, Inc.; p. 32 © GIPhotoStock/ Photo Researchers, Inc.; p. 34 © www.istockphoto.com/Kristian Stensønes; p. 35 Wikimedia Commons; p. 37 © Roger Hutchings/Corbis; p. 38 © BSIP/Photo Researchers, Inc.

Designer: Tahara Anderson; Editor: Nicholas Croce;
Photo Researcher: Cindy Reiman